THE LIBRARY OF FOOD CHAINS AND FOOD WEBS

Food Chains in a
POND HABITAT

ISAAC NADEAU
Photographs by
DWIGHT KUHN

The Rosen Publishing Group's
PowerKids Press™
York

For Emily, full of life—Isaac Nadeau
For Clark—Dwight Kuhn

Published in 2002 by The Rosen Publishing Group, Inc.
29 East 21st Street, New York, NY 10010

Copyright © 2002 by The Rosen Publishing Group, Inc.

All rights reserved. No part of this book may be reproduced in any form without permission in writing from the publisher, except by a reviewer.

First Edition

Book Design: Emily Muschinske
Project Editor: Emily Raabe

All photographs © Dwight Kuhn except pp. 10 (beaver) and 20 (beaver and stump) © CORBIS.

Nadeau, Isaac.
 Food chains in a pond habitat / Isaac Nadeau.
  p. cm.— (The library of food chains and food webs)
 Includes bibliographical references.
 ISBN 0-8239-5763-2 (lib.)
 1. Pond ecology—Juvenile literature. [1. Pond ecology. 2. Food chains (Ecology). 3. Ecology.]
 I. Title. II. Series.
QH541.5.P63 N34 2002   2002-000150
577.63'616—dc21

Manufactured in the United States of America

Contents

1	Food Chains in a Pond	4
2	What Makes a Pond a Pond?	6
3	The Food Makers	8
4	Underwater Herbivores	10
5	It's a Frog Eat Frog World	12
6	Omnivores and Scavengers in the Pond	14
7	The Joy of Rot	16
8	Life Among the Tiny	18
9	A Watery Web of Life	20
10	Exploring the Pond	22
	Glossary	23
	Index	24
	Web Sites	24

Food Chains in a Pond

Would you believe that all of the energy in a pond comes from one place? That place is the sun. The sun is the origin of all of the energy in the food chains in a pond.

If you ever have been to a pond, you have been to a place with food chains. In fact, there are food chains wherever you go. That is because all living things, from plants to people, are part of food chains. A food chain is a way to describe how energy passes from one living thing to the next. Energy is the **fuel** that makes life possible. Plants and animals get their energy from food. Each time a plant or an animal is eaten, a link is added to the food chain. There are many food chains in a pond. Day and night, pond creatures are hard at work finding food and trying to stay alive. All of the beings in a pond need energy to grow and move about.

This green algae, wood frog tadpole, pumpkin seed sunfish, and great blue heron form just one of many food chains in a pond.

What Makes a Pond a Pond?

A habitat is a place where a plant or an animal lives. The pond habitat is a place where plants and animals can find all the comforts of home, including food, shelter, and a safe place to raise their babies.

Water is the most important element of a pond habitat. The water in a pond is usually shallow. Most ponds have a muddy bottom. Besides the water, what really makes a pond a pond are the things that live there. Every plant and animal in a pond habitat has **adapted** to life in the pond. Some animals, such as bluegills and other fish, have gills, which they use to breathe underwater. Some plants, such as water lilies, have flat leaves that float on the surface of the pond to catch sunlight. Many animals, including frogs, turtles, and salamanders, bury themselves in the mud at the bottom of the pond to **hibernate** over the winter.

Bottom Right: The fuzzy things coming out of this young spotted salamander's sides are gills. Baby salamanders use their gills to breathe underwater.

Right: This green frog does not have the gills it had as a tadpole. As an adult, it uses lungs to breathe.

Below: This female spotted salamander is laying her eggs.

The Food Makers

Plants are the first link in any pond food chain. All life in a pond depends on the energy that plants gather from the sun. Through a process called **photosynthesis**, plants use sunlight, water, and air to produce sugar. Plants use the energy from sugar to grow leaves, flowers, fruits, and seeds and to pull **nutrients** up through their roots. Whatever energy is left over is stored in the plants' bodies. When a tadpole or other animal eats the plant, it gets this leftover energy. Plants are called **producers** because they produce the food needed by the animals in a food chain.

Some of the most common plants in a pond are called **algae**. Algae are very tiny, but there are so many of them that they often turn a pond green. As simple as they are, algae may be the most important producers in the pond. Many animals get their food from algae or eat animals that eat algae.

Of all the wonderful things that live in and around a pond, only the plants are able to use sunlight as food.

Underwater Herbivores

Beavers are also pond herbivores. Unlike water fleas, beavers spend only part of their time under water. Beavers are also much larger than most of the animals that live in the pond. Beavers build their homes, called lodges, out of the branches of trees that grow near the pond. They feed on the bark of willow, alder, and other trees.

Herbivores are animals that eat only plants. Herbivores are the second link in a pond food chain. The water flea is a common herbivore. This tiny animal uses its many legs to swim and to sweep algae into its mouth. The water flea produces thousands of eggs. Only a few of these eggs grow up to be adult water fleas. Most of them become food for other animals. Snails are another common herbivore in the pond habitat. There are many kinds of snails in a pond.

Many pond herbivores, such as tadpoles, are herbivores early in their lives but grow up to be **predators**, or animals that hunt other animals for food.

Right: *Snails use their mouths to scrape algae off rocks and the stems of plants.*

Left: *You can see eggs inside the body of this water flea.*

It's a Frog Eat Frog World

Few habitats are home to more predators than a pond. Even the hunters are being hunted in the pond! Animals that eat other animals are called **carnivores**. Carnivores are the third link in a pond food chain. They also may be the fourth or even fifth link, because sometimes carnivores eat other carnivores. Most fish are carnivores. Many of the insects that live in ponds are also carnivores. Dragonflies are some of the deadliest carnivores in the pond. Dragonflies are carnivores even when they are just babies, or nymphs. Dragonfly nymphs eat water mites, insects, and tadpoles. As adults, dragonflies catch flying insects in midair. Tadpoles that grow up to be frogs without being eaten by dragonflies have a chance to get back at the dragonflies. Adult frogs eat both dragonfly nymphs and adult dragonflies.

Frogs also eat worms, snails, and other insects. Frogs even eat tadpoles!

They may be small, but this dragonfly (top left), bullfrog (top right), and giant water bug (bottom left) are all fearsome predators to tiny pond dwellers.

Omnivores and Scavengers in the Pond

For the **omnivores** in a pond, the world is like one big bowl of food. That's because omnivores eat all kinds of food, both plants and animals. Turtles are omnivores. Often a turtle will even eat the body of a dead animal. Animals that eat dead animals are called **scavengers**. Flatworms are also scavengers. Flatworms are tiny animals that wriggle about on the pond's muddy bottom in search of food.

Any animal would be in big trouble if its food source disappeared. For example, without algae, the water flea would have nothing to eat. Omnivores and scavengers are lucky because they eat many different things. If one food is gone, they can find other kinds. Omnivores are part of many, many food chains in a pond.

Right: *Crayfish are omnivores that have powerful claws to help them catch food.*

Top Right: *This tiny flatworm is eating a mosquito larva.*

Top Left: *Turtles, like the painted turtle shown here, eat fish, frogs, algae, seeds, and many other foods.*

The Joy of Rot

Without the decomposers, all of the nutrients in a pond food chain would be stuck in the bodies of dead plants and animals. There would be no nutrients left to go back to the roots of the plants. Without these nutrients, the plants could not photosynthesize, and there would be no life in the pond.

Nothing on Earth can live without having to die someday. This is true in a pond, too. The food chain does not come to an end, however, when an animal dies. There is one last, important step for the energy to take. The last link in a pond food chain is formed by the **decomposers**. These are the tiny creatures that live in the mud and float in the water, waiting to help plants and animals rot. There are two main types of decomposers in a pond habitat. These are **bacteria** and **fungi**. Bacteria and fungi are everywhere in the pond. They feed on dead things. When they are done eating, all that is left of the plant or animal's body is the nutrients.

The water mold growing on this dead salamander is only one of the decomposers hard at work breaking down its body.

Life Among the Tiny

   Some of the most amazing creatures in the pond are so small they can't be seen without a microscope. An entire food chain of these creatures, including producers, herbivores, carnivores, omnivores, and decomposers, could fit inside 1 tablespoon (1.5 cl) of pond water! In this world of tiny creatures, the giant is the **amoeba**. Amoebas are about the size of pinheads, but they are still 1,000 times larger than many of the bacteria that live in the pond.

   Another interesting and tiny pond creature is called the water bear. The water bear looks like a real bear, but it has eight legs and is only about the size of an amoeba. Water bears crawl along the leaves of underwater plants and suck the juices out of their **cells**.

Above: *This amoeba is smaller than the head of a pin!*

A Watery Web of Life

Every living thing in a pond is connected to every other living thing. This pattern is called a food web. The arrows in this web point to the creature that is getting the energy. The decomposers, shown at bottom right, will help to break down the body of every creature in the food web after it dies.

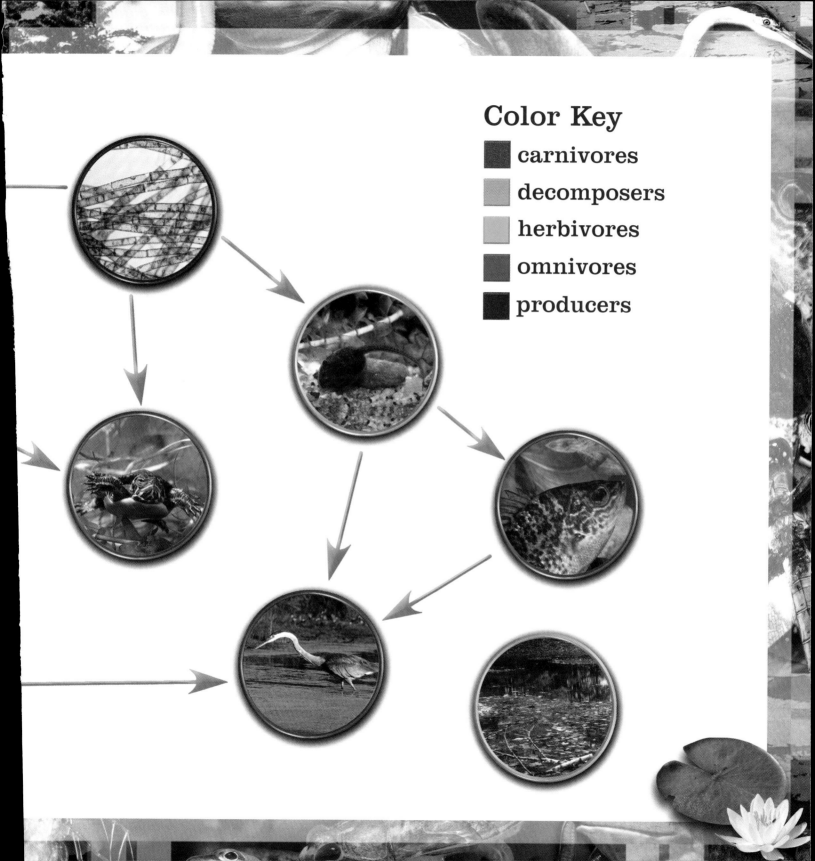

Exploring the Pond

If you do visit a pond, remember to bring an adult with you, and remember that a pond is someone's home! Be sure to return whatever animals you take out of the pond.

The best way to learn about the web of life in a pond habitat is to see it for yourself. There are many ways to explore the life in a pond. With a dipping net, a bucket, and a magnifying glass, you can discover all kinds of interesting things.

You also can try to create a pond in a jar at home. Take a jar of pond water home and see how many different kinds of plants and animals you find. If you use a magnifying glass or microscope, you will see more plants and animals than you can count. Try drawing the different kinds of living things you see. There are always new things to learn about life in a pond.

Glossary

adapted (uh-DAPT-id) Changed to fit conditions.
algae (AL-jee) Plants without roots or stems that usually live in water.
amoeba (uh-MEE-buh) A tiny animal that lives in water and uses jelly-like motions to move.
bacteria (bak-TEER-ee-uh) Tiny living things that are seen with a microscope.
carnivores (KAR-nih-vorz) Animals that eat other animals for food.
cells (SELZ) Tiny units that make up all living things.
decomposers (dee-kuhm-POH-zerz) Organisms, such as fungi, that break down the bodies of dead plants and animals.
fuel (FYOOL) Something used to make energy, warmth, or power.
fungi (FUN-gy) Living things that are like plants but without flowers, leaves, or green coloring.
herbivores (ER-bih-vorz) Animals that eat plants.
hibernate (HY-bur-nayt) To spend the winter sleeping or resting.
nutrients (NOO-tree-intz) Things a living thing needs to live and grow.
omnivores (AHM-nih-vorz) Animals that eat both plants and animals for food.
photosynthesis (foh-toh-SIN-thuh-sis) The process in which leaves use energy from sunlight, gases from air, and water from soil to make food.
predators (PREH-duh-terz) Animals that kill other animals for food.
producers (pruh-DOO-serz) Plants that use sunlight to make energy and are eaten by animals higher in the food chain.
scavengers (SKA-ven-jerz) Animals that feed on dead animals.

Index

A
algae, 8, 14
amoeba, 18

B
bacteria, 16, 18

C
carnivores, 12, 18

D
decomposers, 16, 18
dragonflies, 12

E
energy, 4, 8, 16

F
frogs, 12
fungi, 16

H
habitat(s), 6, 12
herbivores, 10, 18

I
insects, 12

N
nutrients, 8, 16

O
omnivores, 14, 18

P
photosynthesis, 8
predators, 10, 12
producers, 8, 18

S
scavengers, 14

T
tadpole(s), 8, 12

W
water, 6, 8, 16, 18, 22
web of life, 22

Web Sites

To learn more about pond habitats, check out these Web sites:

http://web.ukonline.co.uk/conker/pond-dip/
http://octopus.gma.org/turtles/pond.html
http://mbgnet.mobot.org/fresh/lakes/index.htm